COUNTDOWN TO SPACE

COUNTDOWN TO SPACE

URANUS—
The Seventh Planet

Michael D. Cole

Series Advisors:
Marianne J. Dyson
Former NASA Flight Controller
and
Gregory L. Vogt, Ed. D.
NASA Aerospace Educational Specialist

Enslow Publishers, Inc.

40 Industrial Road	PO Box 38
Box 398	Aldershot
Berkeley Heights, NJ 07922	Hants GU12 6BP
USA	UK

http://www.enslow.com

Library of Congress Cataloging-in-Publication Data

Cole, Michael D.
    Uranus : the seventh planet / by Michael D. Cole.
        p. cm. — (Countdown to space)
    Summary: Explores the seventh planet of our solar system, its composition, terrain, and unique features, as well as historical and current scientific explorations.
    Includes bibliographical references and index.
    ISBN 0-7660-1952-7
    1. Uranus (Planet)—Juvenile literature. [1. Uranus (Planet)] I. Title. II. Series.
    QB681 .C65 2002
    523.47—dc21
                                       2002001186

Printed in the United States of America

10 9 8 7 6 5 4 3 2 1

To Our Readers: We have done our best to make sure all Internet Addresses in this book were active and appropriate when we went to press. However, the author and the publisher have no control over and assume no liability for the material available on those Internet sites or on other Web sites they may link to. Any comments or suggestions can be sent by e-mail to comments@enslow.com or to the address on the back cover.

Photo Credits: William Sauts Bock, p. 17; Enslow Publishers, Inc., pp. 7, 8; © 2001 Calvin J. Hamilton, pp. 10, 27; Erich Karkoschka (University of Arizona) and NASA, pp. 22, 23, 35; Lunar and Planetary Institute (LPI), pp. 20, 21, 39; NASA, pp. 4, 29, 30, 31; NASA/JPL, pp. 25, 33; Royal Astronomical Observatory/ European Space Agency, p. 13; Paul Schenk (LPI), p. 24; USGS (Miranda), p. 25; H. Weaver (ARC), HST Comet Hyakutake Observing Team, and NASA, p. 15.

Cover Photo: NASA (foreground); Raghvendra Sahai and John Trauger (JPL), the WFPC2 science team, NASA, and AURA/STScI (background).

CONTENTS

Uranus—the seventh planet from the Sun.

1

Spring on an Alien World

The Hubble Space Telescope orbited high above Earth. From its position in space, the orbiting telescope took pictures that were not blurred by Earth's atmosphere. Its pictures of planets, stars, galaxies, and other space objects were clearer than the pictures taken through any telescope in an observatory on Earth. Seven times between 1994 and 1998, the powerful telescope turned its instruments toward the faraway planet Uranus.

Many pictures were taken of the planet and its faint system of rings during those years. In 1999, scientists used a computer to piece the pictures together, one after the other, until the stream of pictures looked like a movie. Uranus was thought to be a featureless, dull-looking world. But the results of the computer-animated pictures were startling. There were massive spring storms on Uranus!

The storms raging on the distant planet were nothing like spring storms on Earth. The storm clouds on Uranus were not producing rain or flashes of lightning. Unlike clouds on Earth, which are made of water vapor, the clouds on Uranus are made of hydrogen and helium gases. The images showed that each of the storms was large enough to cover the United States from Kansas to New York. And the temperature in these spring storms was –300°F (–184°C). The storms on Uranus were developing because the planet's northern hemisphere was starting to come out of a cold and very long winter.[1]

We experience seasons on Earth because Earth's north and south poles are tilted slightly with respect to the Sun's north and south poles. This angle leaves one part of Earth tilted more directly toward the Sun during one season of the year, and slightly away from the Sun during another season.

The north and south poles of Uranus, however, are tilted almost completely sideways. Other planets in the solar system spin through space like a top as they orbit the Sun. But Uranus rolls along its equator, traveling through space like a bowling ball rolling down a bowling lane.

With its poles tipped in this fashion, the south pole and much of the southern hemisphere of Uranus spend the entire summer facing the Sun. The north pole and

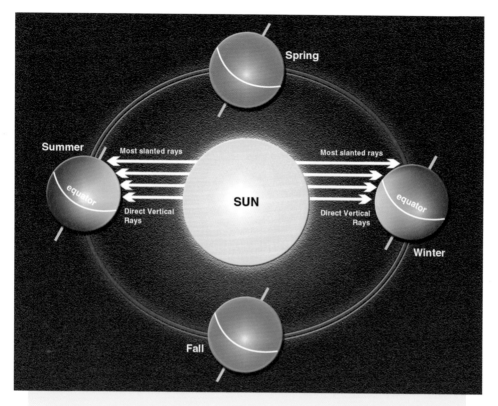

Earth's seasons are caused by the planet's tilt. When the Northern Hemisphere is tilting toward the Sun, it is summer. When the Northern Hemisphere is tilting away from the Sun, it is winter.

northern hemisphere spend that time facing away from the Sun. Although a "day" on Uranus occurs every time the planet rotates once on its axis, these days near the planet's south pole are without a sunrise or sunset. With the planet's south pole facing almost directly toward the Sun, the Sun's light shines day after day in the southern hemisphere. The darkness at Uranus's north pole is complete.

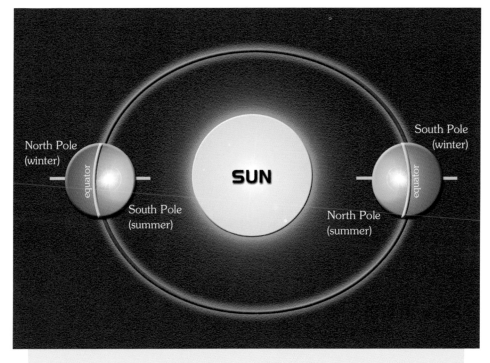

North Pole
(winter)

equator

SUN

South Pole
(summer)

South Pole
(winter)

North Pole
(summer)

equator

Uranus rolls along its equator as it orbits the Sun. When the south pole of the planet faces the Sun, it is summer on that hemisphere. There is complete darkness on the north pole at this time.

By the late 1990s, Uranus had come around in its orbit to a point where the Sun had begun to shine on parts of both the northern and southern hemispheres of the planet. It was springtime in Uranus's northern hemisphere, and the storms were a result of that hemisphere starting to warm up.

Scientists had been observing Uranus in telescopes for two hundred years. Why had no one observed these storms before? Dr. Heidi Hammel, a planetary scientist studying the Hubble Space Telescope images at the Massachusetts Institute of Technology (MIT), had an answer.

"No one has ever seen this view [of Uranus] in the modern era of astronomy," she said, "because of the long year of Uranus—more than eighty-four Earth years."[2]

Because it takes eighty-four Earth years for Uranus to complete one orbit around the Sun, the seasons on Uranus are very long. The planet's northern hemisphere had spent the last twenty years facing away from the Sun. Scientists had spent a long time looking at the planet's southern hemisphere. The *Voyager 2* spacecraft, the only spacecraft to pass Uranus, saw no such storm activity when it passed the planet in 1986.

"It just looked like a seamless blue tennis ball," said scientist Ellis Minor of NASA's Jet Propulsion Laboratory (JPL) in Pasadena, California.[3]

Wes Lockwood, an astronomer at Lowell Observatory in Flagstaff, Arizona, agreed. "It was thought to be dull because it looked dull once and no one looked at it again."[4]

Only early observers of Uranus, back in the 1780s, had ever described bands of colorful clouds around the planet. It had been assumed that these observations were more likely the result of imperfect focusing of light in early telescopes. The images from the Hubble Space Telescope raise the possibility that these early

When Voyager 2 photographed Uranus in 1986, scientists said it looked like a seamless blue tennis ball.

astronomers were seeing similar storms in the planet's atmosphere.

Many astronomers believe the clouds seen by the Hubble Space Telescope were formed after the Sun began to heat the hemisphere that had been kept in frigid darkness for so long. Another possibility is that the clouds existed on the planet's northern hemisphere and were hidden from scientists because they had been on the far side of the planet.

"Nobody knows for sure," Hammel added, "but it would be strange if the two hemispheres were really so different."[5]

The unexpected storms were not the only strange thing the computerized movie showed. The faint ring system that exists around Uranus was wobbling like an unbalanced wagon wheel as the planet rotated. It suddenly seemed that the seventh planet from the Sun was neither featureless nor dull.

"This movie is completely changing people's understanding of Uranus," Hammel said.[6]

Uranus today is surprising scientists, forcing them to look at the planet from a new perspective. The challenge to understand Uranus began more than two hundred years ago with its discovery, which was also a surprise.

2

A Seventh Planet?

The Sun and the Moon were not the only objects in the sky that caught the attention of people in ancient times. Thousands of years ago, early civilizations also identified five wanderers, or planets, that moved across the background of stars. The planets Mercury, Venus, Mars, Jupiter, and Saturn were visible in the night sky with the naked eye. People from many early civilizations saw these bright planets and assigned them the names of various gods from their cultures.

Polish astronomer Nicolaus Copernicus (1473–1543) believed that Earth revolved around the Sun. Most astronomers at this time, however, thought that Earth was the center of the universe and that bodies in the solar system revolved around Earth. From 1590 to about

1600, the observations of Danish astronomer Tycho Brahe and the calculations of German mathematician Johannes Kepler helped prove that the known planets, as well as Earth, were in orbit around the Sun. Our early scientific picture of the universe took shape. In this picture, a solar system of six planets, including Earth, orbited the Sun.

On the evening of March 13, 1781, that picture changed.

Discovering a "New" Planet

British astronomer William Herschel was observing an area of the constellation Gemini in his telescope. Herschel had built the telescope himself. It was a reflecting telescope, which means there was a mirror at the bottom of the telescope that collected light and reflected it back to another smaller mirror near the top. The smaller mirror was set at an angle that focused the light into an eyepiece attached near the top of the telescope. Herschel was using this and other telescopes to count stars in certain areas of the sky. He

William Herschel was a British astronomer who built his own telescope. He used it to search the night sky.

planned to use these observations to make maps of the sky for other astronomers. It is this project that brought his attention to a small area of Gemini that evening.[1]

"On Tuesday, the 13th of March, between ten and eleven in the evening, while I was examining the small stars in the neighborhood of H Geminorum [a star in the constellation Gemini], I perceived one that appeared visibly larger than the rest," Herschel later wrote. He was surprised by its size and brightness and wanted to compare it with two known stars in that part of the sky. Herschel quickly compared the object with the two stars, "and finding it so much larger than either of them," he wrote, "suspected it to be a comet."[2]

A comet is a chunk of frozen gases and dust that moves in very elliptical orbits around the Sun. As a comet gets nearer the Sun, the Sun's energy heats the comet's surface, releasing some of the gas and dust to form a tail behind the comet. Comets stay mostly invisible until this heating by the Sun begins. The tail grows longer as the comet draws nearer the Sun, making it more visible from Earth.

Herschel was pleased with the possibility that he had discovered a comet. He believed it was a comet because its diameter increased when he increased the magnification of his telescope. This does not happen with stars. Because of the enormous distance of the stars from Earth, increasing magnification does not increase the diameter of a star's light.

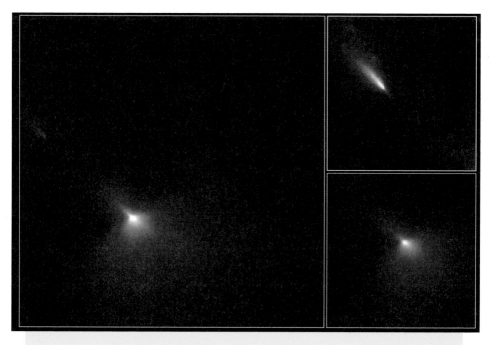

William Herschel initially thought that the new planet could be a comet. These photos show a comet called Hyakutake on March 27, 1996. They were taken through a red filter on the Hubble Space Telescope.

Herschel observed the object again on the night of March 17, 1781. He carefully checked its current position in relation to the background of stars. Herschel then compared it to the position it held on the night of March 13. The object had indeed moved.

The following day Herschel reported his discovery by letter to the Astronomer Royal of England, Nevil Maskelyne. Maskelyne made several observations of the object on his own. To Herschel's surprise, Maskelyne was not convinced that the object was a comet.

Maskelyne explained in a letter to Herschel that he had not observed any fuzziness around the object, and no tail was yet visible. If Maskelyne had observed these two characteristics, he would have confirmed the object as a comet. He wondered instead if the object might be a new planet.[3]

Astronomers in other countries began detailed observations of Herschel's mystery object. One of them was German astronomer Johann Elert Bode. Once a sufficient number of observations of a moving object in space have been made, mathematicians can calculate the object's orbit. Bode made these calculations for the object Herschel had discovered. By November 1781, Bode's calculations proved that the object was a planet. Bode himself was amazed with the calculations. They showed that the newly discovered planet orbited the Sun at a distance twice as far as Saturn. Almost overnight, Herschel's discovery had doubled the size of the known solar system.[4]

Herschel and the Naming of a Planet

As the discoverer of the first new planet since antiquity, Herschel became famous immediately. England's King George III awarded him a royal grant to continue his astronomical research projects. He was knighted by the king in 1816, becoming Sir William Herschel. Herschel was also given the honor of naming the new planet. In appreciation for the grant from his king, Herschel proposed naming the planet Georgium Sidus—George's Star.

Astronomers outside of England were not pleased at the idea of naming the new planet after a British king. For many years, they instead referred to the planet as "Herschel."

Bode was the first astronomer to suggest naming the planet Uranus. He believed astronomers should follow the tradition of naming the planets after characters from classical mythology. The planet Jupiter is the fifth planet

Uranus was the father of Saturn in Roman mythology. He was also known as the ruler of the sky in Greek mythology.

from the Sun, and Saturn is the sixth. In Roman mythology, Saturn was the father of Jupiter, and so, Bode reasoned, the new planet should be named after the father of Saturn—Uranus.[5]

Astronomers called the planet Georgium Sidus, Herschel, and some other names for more than forty years. It was not until after Herschel's death in 1822 that the planet was officially given the name Uranus by the astronomical community.

3

The Distant Seventh Planet

The planet that Herschel discovered in his telescope is about 1.8 billion miles (2.9 billion kilometers) from Earth. Uranus orbits the Sun at a distance nineteen times farther from the Sun than Earth does.

Uranus is not a terrestrial planet, meaning it is not made of rock like Earth, Venus, Mercury, and Mars, the four planets closest to the Sun. It is a giant gas planet— an enormous ball of gas with only a small rocky core at its center. Jupiter, Saturn, and Neptune are also gas giants.

Uranus is more than 31,600 miles (51,000 kilometers) wide, compared to Earth, which is about 8,000 miles (13,000 kilometers) wide. If Uranus were hollow, it could hold sixty-three Earths.

The four gas giant planets—Jupiter, Saturn, Uranus, and Neptune—are shown in size relation to each other. All of these planets are made of gas.

Composition of Uranus

The planet is surrounded by a thick atmosphere of hydrogen and helium gas. The composition of the atmosphere is 82 percent hydrogen, 15 percent helium, 2 percent ammonia, and traces of other gases. This gaseous layer of Uranus is probably about 6,000 miles (9,700 kilometers) deep. At 1.8 billion miles from the Sun, the temperature in Uranus's atmosphere averages a frigid –360°F (–218°C).[1]

Scientists believe that beneath the atmosphere is a layer that consists of a mixture of water, methane, and ammonia molecules. These elements exist in a hot, souplike form that planetary scientists refer to as "ice." But this layer of "ice" is definitely not frozen. It probably exists at a temperature of about 4,200°F (2,300°C). Scientists believe this layer is about 6,000 miles (9,700 kilometers) deep. It is under tremendous pressure from the thousands of miles of gaseous material pressing down from above. The pressure on this material is what makes this layer so hot.[2]

Although the temperature at this layer is great, it is not high enough to turn the hydrogen molecules to a liquid. Temperatures within the planets Jupiter and Saturn create a layer of liquid hydrogen within those planets. The movement of liquid hydrogen creates electricity and additional heat from within those planets. But because Uranus's internal temperature is too low for

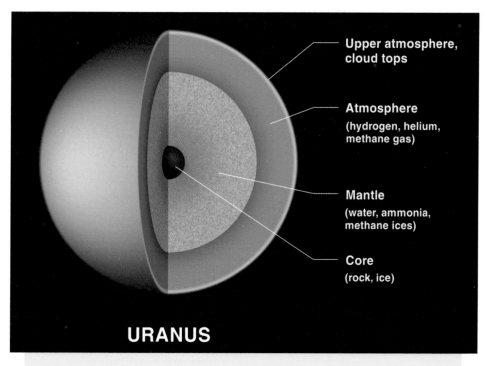

Upper atmosphere, cloud tops

Atmosphere
(hydrogen, helium, methane gas)

Mantle
(water, ammonia, methane ices)

Core
(rock, ice)

URANUS

Scientists believe that the center core of Uranus is made of ice and rock.

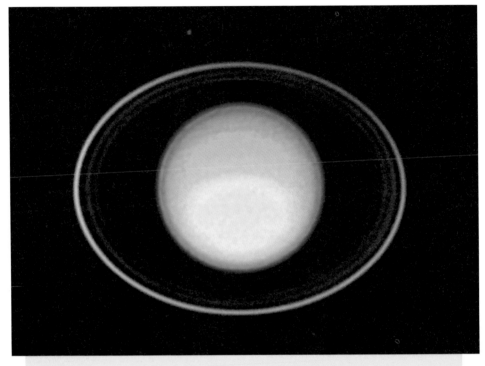

The Hubble Space Telescope took a close look at the atmosphere of Uranus. The red around the planet's edge is a very thin haze at a high altitude. The yellow near the bottom is another hazy layer of the atmosphere. The deepest layer, the blue near the top, shows a clearer atmosphere.

the formation of liquid hydrogen, the planet lacks a significant internal heat source.

Scientists once thought that Uranus's layer of water and ammonia molecules might generate enough electricity to produce the planet's magnetic field. The magnetic field is the area around a planet where the flow of magnetic energy from the planet is stronger than the magnetic energy from the Sun. But data from the *Voyager 2* spacecraft found no evidence that this layer was the source of the magnetic field. Its source is still unknown.[3]

At the center of Uranus is a rocky or molten core that is probably smaller than Earth. Scientists estimate that the core is about 6,000 miles (9,700 kilometers) wide. It is under great pressure from the enormous amount of material surrounding it. The temperature within this core may be more than 12,600°F (7,000°C). Scientists cannot be certain whether this core is solid or molten.

The Rings of Uranus

Uranus is encircled by a series of rings. This system of rings is very different from the wide and complex system that surrounds the planet Saturn. The rings of Uranus are very thin and are made of a dark, dusty material that reflects little light. They are invisible to telescopes on Earth and could be seen and photographed by the *Voyager 2* spacecraft only with special cameras and when backlit by the Sun.

The rings were discovered in 1977 when astronomers observed an event called an occultation. As Uranus passed in front of a star, scientists on

This Hubble Space Telescope image shows the four major rings around Uranus. It also shows ten of the moons that orbit the planet.

Earth used special equipment to measure small changes in the star's light intensity. Before the planet itself passed in front of the star, there were six tiny decreases in the star's light intensity. The drop in light intensity was great as the sphere of Uranus slowly passed in front of the star, directly blocking much of the star's light. After the planet passed by, the scientists' instruments again showed six tiny drops in the star's light intensity. The drops in intensity were identical to the first six. Further observations showed even more identical drops in light intensity. These observations showed scientists that there was a system of at least nine rings around Uranus. *Voyager 2* discovered two additional rings nine years later.[4]

Uranus's Moons

Uranus is also orbited by a number of moons. William Herschel discovered the planet's two largest moons, Titania and Oberon, in 1787. Titania is about 981 miles (1,600 kilometers) wide, while Oberon is 946 miles (1,500 kilometers) wide, making them less than one third

There are five major moons of Uranus: Miranda, Ariel, Umbriel, Titania, and Oberon. They are made of water and other ices, as well as rocky material.

Five Major Moons
All are about 50% water, 20% carbon and nitrogen, 30% rock, and gray in color

Miranda	Ariel	Umbriel	Titania	Oberon
12-mile-deep canyons, −335°F (−204°C)	Brightest moon, newest surface	Dark, pocked with craters	Largest moon, huge fault canyons	Many craters with dark material

Others

Cordelia—only moon inside rings	Juliet	Caliban
Ophelia	Portia	Prospero
Bianca	S/1986	Stephano
Cressida	Belinda	Sycorax
Rosalind	Puck—largest minor moon (95	Setebos
Desdemona	miles [153 kilometers] wide)	

the size of Earth's Moon. Two smaller moons, Ariel and Umbriel, were seen for the first time in 1851, followed by much smaller Miranda in 1948. *Voyager 2*'s visit to Uranus, and further discoveries since then, have brought the current total of known moons orbiting the planet to twenty-one. There may be others still to be discovered.

The enormous distance separating Uranus from Earth made gaining knowledge about the planet, its rings, and its moons very difficult. By the late 1970s, a voyage to this remote world was the only way to learn more about the planet.

URANUS[5]

Age
About 4.5 billion years

Diameter
31,600 miles (51,000 kilometers), nearly 4 times the diameter of Earth

Planetary mass
14½ times as massive as Earth

Distance from the Sun
1.8 billion miles (2.9 billion kilometers), 19 times more distant from the Sun than Earth

Closest passage to Earth
1.5 billion miles (2.4 billion kilometers)

Farthest passage from Earth
1.9 billion miles (3.1 billion kilometers)

Orbital period (year)
84 Earth years

Rotation period (one rotation on its axis)
17 hours, 14 minutes

Temperature
-323°F (-197°C), increases with depth

Composition
Solid or molten rock core (scientists unsure); enormous layer of water, methane, and ammonia "ice;" mostly hydrogen gas atmosphere

Atmospheric composition
82% hydrogen, 15% helium, 2% methane, traces of other gases

Wind speeds
Up to 446 miles (718 kilometers) per hour

Gravity
About 89 percent of Earth's surface gravity in
Uranus's outer atmosphere

Number of known moons
21

Composition of rings
Dust, rock, and rubble

Tilt of axis
97.8 degrees

Brightness of Sun from Uranus
363 times less bright as from Earth

4

A Look at Uranus Up Close

The mission of NASA's *Voyager 2* spacecraft was a tremendous success long before it reached Uranus. The spacecraft was launched from the Kennedy Space Center at Cape Canaveral, Florida, on August 20, 1977. It made many discoveries during flybys of Jupiter, in June 1979, and Saturn, in August 1981.

Voyager 2 arrived at Uranus in January 1986. Over a period of weeks, the spacecraft approached and then passed the planet. *Voyager 2* sent back more than seven thousand images of Uranus, its ring system, and its moons during the mission.

The spacecraft passed within 50,000 miles (80,000 kilometers) of the planet's cloud tops. These close-up images of cloud movement helped scientists determine

After blasting off aboard a rocket, the Voyager 2 *spacecraft arrived at* Uranus *in January 1986.*

how fast Uranus's atmosphere was moving around the planet. The images allowed scientists to record wind speeds of up to 450 miles (720 kilometers) per hour in the upper atmosphere. Other pictures, taken at different wavelengths of light, allowed scientists to see clouds that existed deeper inside the planet's atmosphere.

Other instruments took readings of Uranus's magnetic field. The instruments showed that the magnetic field around Uranus is about the same as the one around Earth. *Voyager 2*'s study of the magnetic field also showed that it was greatly tilted from Uranus's axis of rotation. The north and south magnetic poles of most planets are positioned very close to the north and south poles of their planet's rotation axis. But Uranus's magnetic poles

are tilted nearly 60 degrees away from the planet's axis. No other planet or moon in the known solar system has a tilted magnetic field such as this.[1]

The spacecraft recorded many images of the planet's rings and discovered two additional rings. *Voyager 2*'s

Voyager 2 *took more than seven thousand images of Uranus, its rings, and its moons.*

30

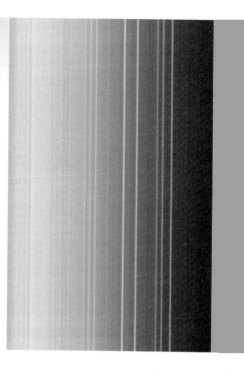

This false-color view of the rings of Uranus was made from images taken by Voyager 2. *All nine known rings are visible here.*

images showed that the rings are very narrow. They are made of small, dark, dusty material. The largest of them is only a few feet in diameter. Some of the rings are incomplete; they do not go all the way around the planet. Scientists think one possible reason for these incomplete rings is that they are not very old. They may be the remains of an object such as a small moon. The object was either shattered by a collision or was torn apart by the gravitational forces of the planet. If the event occurred only hundreds or maybe thousands of years ago, the debris may not have had time to spread all the way around its path around the planet. This is only one possible explanation.

The outermost ring is the Epsilon ring. *Voyager 2* showed that this ring is composed mostly of ice boulders not much larger than beach balls. Most planetary rings found in our solar system contain material of various sizes, from chunks as big as a small mountain to pieces smaller than a dime. The Epsilon ring's lack of small material could be a result of forces exerted on it by two of Uranus's moons.[2]

Uranus's Moons Up Close

Voyager 2 discovered ten additional moons orbiting Uranus. The largest moon discovered by *Voyager 2* was Puck, at 95 miles (153 kilometers) wide. The two smallest moons, Cordelia and Ophelia, were discovered to be less than 20 miles (32 kilometers) wide, making them some of the smallest known moons in the solar system.

These small moons are the two moons that were affecting the planet's Epsilon ring. Cordelia orbits around the inside of the Epsilon ring, while Ophelia orbits around the outside of the ring. Scientists discovered that the gravitational effects of the two moons appeared to be keeping the ring narrow and sharp-edged. It is as if the gravity of the two moons is keeping the ring's particles tidy and neat. Scientists also believe the gravitational effects of the two moons are also sweeping smaller particles, anything less than about two feet in diameter, out of the ring. These smaller particles may be coming to rest on the two moons, or are possibly being sent inward toward the planet.[3]

No such moons were found orbiting near any of the other rings. So the reason for the narrowness of the other rings remains unknown.

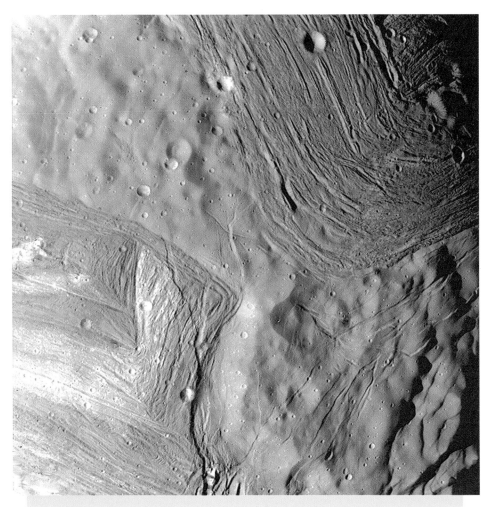

Voyager 2 *took this image of the surface of Miranda. The surface features surprised scientists. The moon has valleys, faults, canyons, and impact craters.*

The moon studied most closely by *Voyager 2* was Miranda. Miranda was the great surprise of the Uranus flyby. The small moon showed the most varied landscape of any moon yet found in the solar system. Its surface showed valleys, fractures, faults, cliffs, canyons, plateaus, and craters. Some of the canyons on Miranda are 12 miles (19 kilometers) deep. In comparison, Earth's Grand Canyon is only about 1 mile (1.6 kilometers) deep.

Scientists believe the moon suffered a violent impact with another object from space before it had a chance to form a core, mantle, and crust. The collision fractured the moon into giant chunks of rock and ice that remained in orbit and later clumped together once more. This second clumping may have caused the strange landforms now seen, as the buried ice tried to rise toward the surface, and the heavier rock tried to sink toward the core.[4]

Some areas of Miranda show very few craters, while others show many craters. Those areas with many craters are older surfaces. More craters exist on these areas because they have had millions more years to accumulate the impacts from space that form the craters. The areas with fewer craters are younger areas, formed by the more recent upwelling of materials from within the re-formed moon. Fewer craters mean the area has had fewer years to accumulate impacts from space.

Uranus's Extreme Axis

The most unique feature of Uranus is the way its rotation axis is tilted. The planet's axis is tilted by 97.8 degrees. This extreme tilt means that Uranus moves through space on its side. The planet's south pole was directly facing the Sun during the *Voyager 2* flyby in 1986. By 1997, Uranus's eighty-four-year orbit had carried it to a position where its northern hemisphere was beginning to experience spring. It was then that a team of scientists at MIT and the University of Arizona

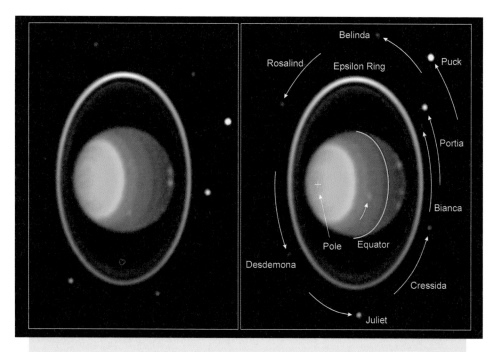

Uranus moves through space on its side. The image on the right was taken ninety minutes after the one on the left. It shows the planet's rotation. Also notice how eight of the planet's moons are included in the images and how their positions changed in the ninety minutes, as shown by the arrows.

used the Hubble Space Telescope to discover the spring storms on Uranus.

"Most people got the impression that Uranus is bland," said project scientist Dr. Erich Karkoschka of the University of Arizona. "As we are now moving towards the end of winter on Uranus, the planet shows remarkable features."[5]

Scientists think it is possible that the planet's extremely tilted axis may cause dramatic seasonal changes in Uranus's atmosphere. But how did the planet come to be tilted this way in the first place?

The Origin of the Planet's Tilt

Scientists believe that an enormous collision occurred as long as 4 billion years ago when Uranus was in its early formation. An object with about the same mass as Earth probably struck Uranus at an area near one of the planet's poles. The blow of this object from space was so violent that it tipped the entire planet over.[6]

Uranus's poles were likely oriented similarly to the Sun's originally. Although there is no real "above" or "below" in space, we think of a planet's north pole as pointing above the orbital plane of the solar system, and the south pole pointing below it. But the tremendous collision suffered by Uranus shifted the position of the planet's poles dramatically.

Observation of the seasonal changes in Uranus's atmosphere by Heidi Hammel and other planetary astronomers may provide more evidence for this theory and help answer other questions about the planet's past and origin.

"Within the next few years, we'll be changing what people read in textbooks and encyclopedias about Uranus," Hammel said. "It's a change at that level of understanding."[7]

Someday in the far future, human beings may visit Uranus or land on one of its moons. If so, it will be a cold and truly alien experience.

5

Experiencing Uranus

After a very long trip through space, the astronauts who visit Uranus will not be getting out to stretch their legs on the planet. There is nothing to walk on.

Because Uranus is a giant gas planet, there would be no attempt to land a spacecraft. Trying to land would only result in the spacecraft sinking farther and farther into the atmosphere. The increasing pressure of all the miles of atmosphere pressing down from above would eventually crush the spacecraft and its occupants. The astronauts will wisely choose a safe orbit around Uranus, or land on one of its moons.

At certain times of the planet's long year, the view of Uranus from a spaceship would, quite honestly, be rather dull. When the planet's poles are facing the Sun, there are practically no atmospheric features visible to the naked eye. At other times, when the planet's equator is turned toward the Sun, storm systems and possibly many bands of clouds may be visible around Uranus.

When the spaceship passes behind to the night side of the planet, one or more of its faint planetary rings may become visible. The Sun from Uranus will appear 363 times less bright than it appears from Earth.[1]

Landing on one of Uranus's many moons could lead to some bizarre experiences. Miranda, with its varied landscape of valleys, cliffs, and canyons would provide many fascinating areas to explore. If the astronauts arrived on Miranda's northern hemisphere when Uranus's north pole is facing the Sun, they would experience no change of night and day on the moon.

This chart shows the approximate sizes of the planets relative to each other. Outward from the Sun (left) are Mercury, Venus, Earth, Mars, Jupiter, Saturn, Uranus, Neptune, and Pluto. Uranus, one of the giant gas planets, does not have a surface for astronauts to land on.

Because the planet's moons orbit an area near the plane of Uranus's equator, the north poles of the moons would be facing toward the Sun as well. If the astronauts landed on the daytime side of Miranda, it would stay daytime for about the next twenty Earth years.

Uranus From Earth

Observing Uranus from Earth is difficult. A good telescope, a detailed star chart, an up-to-date sky map noting the planet's current location, and an experienced observer are all needed to find and observe the planet Uranus. Most telescopes used by amateur astronomers will show only a dull or blurry spot. The size and blurred edges of the spot are all that distinguish it from the surrounding stars.[2]

Ask someone at your school or local library if your community has an astronomy club. Most astronomy clubs host public programs where people can observe through the club members' telescopes. Such a program might be your best chance to look at Uranus and other wondrous objects in space.

Uranus, with the recent discovery of rings, new moons, and violent weather, has proven to be full of surprises. Maybe someday you will be a scientist or astronaut who helps to discover more about this strange and distant planet.

CHAPTER NOTES

Chapter 1. Spring on an Alien World

1. NASA Press Release 99-47, *Huge Spring Storms Rouse Uranus From Winter Hibernation*, March 29, 1999, <http://www.qadas.com/qadas/nasa/nasa-hm/1493.html> (April 17, 2002).

2. Ibid.

3. Robert Miller, "Space scientist brings Uranus, least-loved planet, into focus," *The News-Times Online Edition*, September 3, 2000, <http://www.newstimes.com/archive2000/sep03/lcc.htm> (July 3, 2001).

4. Ibid.

5. Mark Sincell, "Spring on Uranus," *Academic Press Daily inScight Web page*, March 31, 1999, <http://www.academicpress.com/inscight/03311999/graphb.htm> (July 3, 2001).

6. Ibid.

Chapter 2. A Seventh Planet?

1. Patrick Moore, *The Picture History of Astronomy* (New York: Grosset & Dunlap, 1961), pp. 90–91.

2. Michael E. Bakich, *The Cambridge Planetary Handbook* (New York: Cambridge University Press, 2000), p. 264.

3. Ibid., p. 265.

4. Moore, p. 91.

5. Bakich, pp. 266–267.

Chapter 3. The Distant Seventh Planet

1. "Uranus Fact Sheet," *National Space Science Data Center Web Page*, January 9, 2001, <http://nssdc.gsfc.nasa.gov/planetary/factsheet/uranusfact.html> (July 3, 2001).

2. J. Kelly Beatty, Carolyn Collins Petersen, and Andrew Chaikin, eds., *The New Solar System* (Cambridge, Mass.: Sky Publishing Corporation, 1999), pp. 196–197.

3. Ibid., p. 198.

4. Jean Audouze and Guy Israel, eds., *The Cambridge Atlas of Astronomy* (Cambridge, England: Cambridge University Press, 1996), p. 214.

5. Michael E. Bakich, *The Cambridge Planetary Handbook* (New York: Cambridge University Press, 2000), pp. 261–275; "Uranus Fact Sheet."

Chapter 4. A Look at Uranus Up Close

1. Jean Audouze and Guy Israel, eds., *The Cambridge Atlas of Astronomy* (Cambridge, England: Cambridge University Press, 1996), p. 213.

2. Michael E. Bakich, *The Cambridge Planetary Handbook* (New York: Cambridge University Press, 2000), p. 273.

3. Ibid.

4. "Miranda and Ariel," *Lunar and Planetary Institute Web Page*, n.d., <http://www.lpi.usra.edu/pub/research/outerp/usat.html> (July 3, 2001).

5. "Hubble Spots Northern Hemisphere Clouds on Uranus," *ARVAL's Gallery Web Page*, November 23, 1997, <http://www.arval.org.ve/uranus.htm> (April 19, 2002).

6. Bakich, p. 268.

7. Robert Miller, "Space scientist brings Uranus, least-loved planet, into focus," *The News-Times Online Edition*, September 3, 2000, <http://www.newstimes.com/archive2000/sep03/lcc.htm> (July 3, 2001).

Chapter 5. Experiencing Uranus

1. Michael E. Bakich, *The Cambridge Planetary Handbook* (New York: Cambridge University Press, 2000), p. 269.

2. Rick Shaffer, *Your Guide to the Sky* (Chicago: Lowell House, 1999), p. 89.

GLOSSARY

atmosphere—The layers of gases surrounding an object in space.

comet—A celestial body that travels in a huge elliptical orbit. When orbiting near the Sun, it develops a long tail that points away from the Sun.

constellation—A pattern or arrangement of stars in a given area of the sky. There are eighty-eight recognized constellations, each with its own name such as Orion or Leo.

crater—A bowl-shaped area created by the impact of another object from space.

flyby mission—A mission in which a spacecraft makes its observations as it passes a planet or other object in space. Many early space missions were designed to fly by, not to orbit or land on, the object they were sent to study.

Hubble Space Telescope—An orbiting observatory equipped with a very powerful telescope. It has viewed objects up to 13 billion light-years away.

magnetic field—The region around a star, planet, or moon where forces due to the electrical current within that body can be detected.

occultation—The passage of one space object, usually a star, behind another of larger size as seen from Earth.

planet—Large bodies that orbit stars.

reflecting telescope—A telescope that uses a mirror to collect and focus light onto another smaller mirror, which angles the collected light through an eyepiece to the viewer.

ring—Large collections of particles, ranging in size from dust to small asteroids, that orbit a planet.

solar system—The Sun and everything held in its gravitational field, including planets, asteroids, and comets.

FURTHER READING

Books

Hunt, Garry E., and Patrick Moore. *Atlas of Uranus.* New York: Cambridge University Press, 1989.

Kerrod, Robin. *Uranus, Neptune, and Pluto.* Minneapolis, Minn.: Lerner Books, 2000.

Miner, Ellis D. *Uranus: The Planet, Rings and Satellites.* New York: John Wiley & Sons, 1998.

Vogt, Gregory. *Jupiter, Saturn, Uranus, and Neptune.* Chatham, N.J.: Raintree Steck-Vaughn, 2000.

Internet Addresses

Arnett, Bill. "Uranus." *Nine Planets.* September 2000. <http://www.nineplanets.org/uranus.html>.

Hamilton, Calvin J. "Uranus." *Views of the Solar System.* n.d. <http://www.solarviews.com/eng/uranus.htm>.

"Fact Sheet: Uranus Science Summary." *Voyager Projects: Uranus.* May 24, 1995. <http://vraptor.jpl.nasa.gov/voyager/vgrur_ fs.html>.

INDEX

47